OUR
PATCHWORK PLANET

THE STORY OF PLATE TECTONICS

BY
HELEN RONEY SATTLER

ILLUSTRATED BY
GIULIO MAESTRO

AND WITH PHOTOGRAPHS

LOTHROP, LEE & SHEPARD BOOKS   NEW YORK

This book is dedicated to the memory of
Helen Roney Sattler
on behalf of her many grateful readers
–THE EDITORS–

ACKNOWLEDGMENTS

With special thanks and appreciation to Jim Lawson, chief geophysicist at the Oklahoma Geological Survey Observatory, near Leonard, Oklahoma, for reading the complete manuscript and adding many valuable pieces of information, and for checking the illustrations for accuracy. In addition to those cited in the "For Further Reading" section, I am indebted to the following writers and scientists for information gleaned from their papers and articles: Philip Abelson, George Alexander, Alan Anderson, Jr., Don Anderson, Stephanie Bernard, Peter Bird, Jeremy Bloxhan, Shannon Brownlee, Stephen Brush, Juan Carlos Castilla, D. Comte, Vincent Courtillot, Donald DePaolo, Wayne Fields, Jean Francheteau, Bruce Frisch, Wilber Garret, Ian Gass, David Gubbins, Egill Hauksson, Thomas Heaton, T. A. Heppenheimer, Randall Hyman, Raymond Jeanloz, Joseph Judge, Ruth Kalamarides, Gregory Lyzenga, Ken MacDonald, Charles Mann, Stephen Maran, Peter Molnar, Cathryn R. Newton, Celia Nyamweru, R. Pelton, Ronald Schiller, Christopher Scholz, Ramond Siever, George Stanley, Joann Stock, Stuart Ross Taylor, Gregory Vink, William Zinsmeister, Mark and D. Zoback.

Photo credits: p. 4 *(top)* Victor Englebert, *(bottom)* Philippe Bourseiller/Sygma; p. 5 © 1989, Comstock, Inc.; p. 7 transparency #K12009, courtesy of Department of Library Services, American Museum of Natural History; pp. 10, 11, 39, 41 courtesy of National Geographical Data Center; p. 18 *(left) courtesy of* NOAA/NGDC, *(right)* U.S. Geological Survey (D. Perkins); p. 24 Pierre Vauthey/Sygma; pp. 27, 33 courtesy of University of Colorado; p. 34 *(left)* © 1992, Comstock, Inc., *(right)* Sygma; p. 36 Reinsurance Company, Munich, Germany; p. 38 R. T. Holcomb, Hawaii Volcano Observatory, courtesy of U.S. Geological Survey.
Cover photo courtesy of National Oceanic and Atmospheric Administration, National Geographical Data Center. Seismogram p. 17 courtesy of Oklahoma Geological Survey Observatory.

Library of Congress Cataloging in Publication Data   Sattler, Helen Roney. Our patchwork planet: the story of plate tectonics / by Helen Roney Sattler; illustrations by Giulio Maestro.
    p.   cm. Includes bibliographical references. Summary: Discusses plate tectonics and the forces that cause motion and change in our planet. ISBN 0-688-09312-4. — ISBN 0-688-09313-2 (lib. bdg.) 1. Plate tectonics — Juvenile literature. [1. Plate tectonics. 2. Continental drift. 3. Geology.] I. Maestro, Giulio, ill. II. Title. QE511.4.S28 1991 551.1'36 — dc20 90-32623 CIP AC

CONTENTS

UNLOCKING EARTH'S MYSTERIES

LONG AGO, PEOPLE THOUGHT THE EARTH NEVER CHANGED. They thought that our world was exactly like it had always been. But when scientists realized that they could not explain why certain things were found in places where they should not be, they began to wonder why. Fossils of marine animals were found in rocks at the top of the Himalaya Mountains. Long rows of sand and gravel showed that a glacier, or ice sheet, had once stretched across the middle of the Sahara Desert. Fossils of trees, ferns, and plant-eating dinosaurs were found in Antarctica.

But Antarctica is much too cold for most plants or plant-eating animals to live there. A glacier could not possibly leave deposits in the middle of a hot desert. Nor could marine animals swim to the top of the world's highest mountains — not if the continents had always been exactly where they are today.

It became evident that something about the world had changed, but what? how? when? and why? Scientists began to search for answers. Just as police detectives gather clues to figure out what went on at the scene of a crime, scientists began to reconstruct the history of Earth from clues.

◄ The layer of Earth that is now the Sahara Desert *(top)* was once covered with glaciers and icebergs, the way Antarctica *(bottom)* is today.

Fossil evidence has also shown that sea creatures such as this starfish, as well as ferns, trees, and plant-eating animals, thrived in Antarctica forty million years ago. ▶

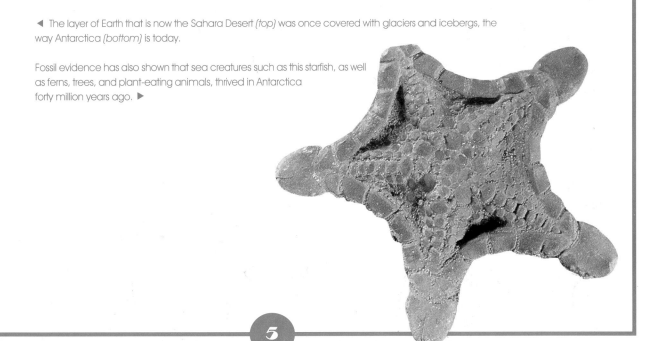

Geologists noticed that the west side of Africa seems to fit against the east side of South America like the pieces of a jigsaw puzzle. They suspected that these continents were once joined together. To test this idea, they studied the formations and composition of coastal rock on both continents and discovered that they match closely. They also found fossils of the same kinds of animals on both continents.

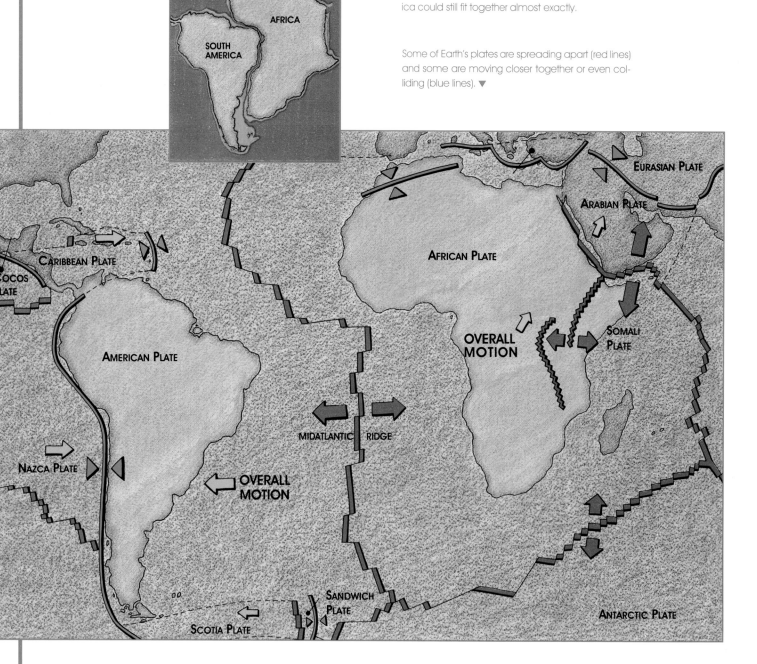

◀ At the edges of their continental shelves, about 3,000 feet below sea level, Africa and South America could still fit together almost exactly.

Some of Earth's plates are spreading apart (red lines) and some are moving closer together or even colliding (blue lines). ▼

SOUTH AMERICA

AFRICA

EURASIAN PLATE

ARABIAN PLATE

CARIBBEAN PLATE

COCOS PLATE

AFRICAN PLATE

SOMALI PLATE

OVERALL MOTION

AMERICAN PLATE

NAZCA PLATE

MIDATLANTIC RIDGE

OVERALL MOTION

SANDWICH PLATE

SCOTIA PLATE

ANTARCTIC PLATE

Although Earth's crust is very thin compared to Earth's size, the layers of crust plainly visible in the monumental Grand Canyon only suggest its depth. The outer layer of the globe averages about 25 miles, or 40 kilometers (km), thickness beneath the continents and about 4.5 miles (7.2 km) below the ocean. If the planet were the size of a peach, its crust would be no thicker than the peach's skin.

These clues showed what had happened: Continents had moved. But how and why remained a mystery. Geophysicists drilled deep holes into Earth in their search for answers. They also studied the walls of the Grand Canyon. This mile-deep gash, cut into the planet's crust by the Colorado River, exposes layer after layer of rock.

Scientists found out the age of each layer by measuring the amount of radioactive materials in the rocks. They discovered that Earth is much older than they thought. The rock in the bottom layer of the canyon is about 2 billion years old. Even older rock was found in Canada, India, and Greenland. Scientists think that our world is probably 4.6 billion years old. It has changed many, many times over that period.

Continents and islands have been coming together, splitting apart, folding, sinking, rising, and rotating for billions of years. They seem to be propelled by forces from deep within Earth. Scientists have searched for the source of these forces.

Scientists can learn a lot about the center of the Earth from earthquakes. Several earthquakes happen every day somewhere on Earth. Their shock waves, which are called seismic waves, travel all the way through Earth. Seismographs stationed around the world record and measure them. Studies of seismic waves show that below the thin, solid crust there is an 1,800-mile-deep (2,900 km) layer called the mantle. Below the mantle is Earth's core, which may be as hot as or hotter than the surface of the sun.

The mantle is made of three layers. The bottom layer, the mesosphere, sits on top of the core and is very hot. The middle layer is called the asthenosphere, from the Greek word meaning "weak." The rock in this layer is hot enough to be soft and pliable like Silly Putty. It can be pushed in and stretched out. If you leave a ball of Silly Putty sitting on a table, it will flatten under its own weight. Rock in the asthenosphere flows the same way.

Sound waves carry better in water than in air. Seismic waves slow down when they go from solid rock into semimelted rock. The waves go even slower when they pass through rock that is melted and has become liquid. They also change direction when they go through the Earth's outer core.

Mesosphere comes from the Greek words for "middle" and "ball"; *asthenosphere* from "weak" and "ball"; *lithosphere* from "stony" and "ball."

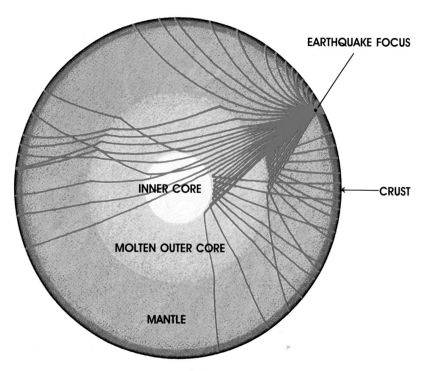

EARTHQUAKE FOCUS

CRUST

INNER CORE

MOLTEN OUTER CORE

MANTLE

PATHS OF SEISMIC WAVES

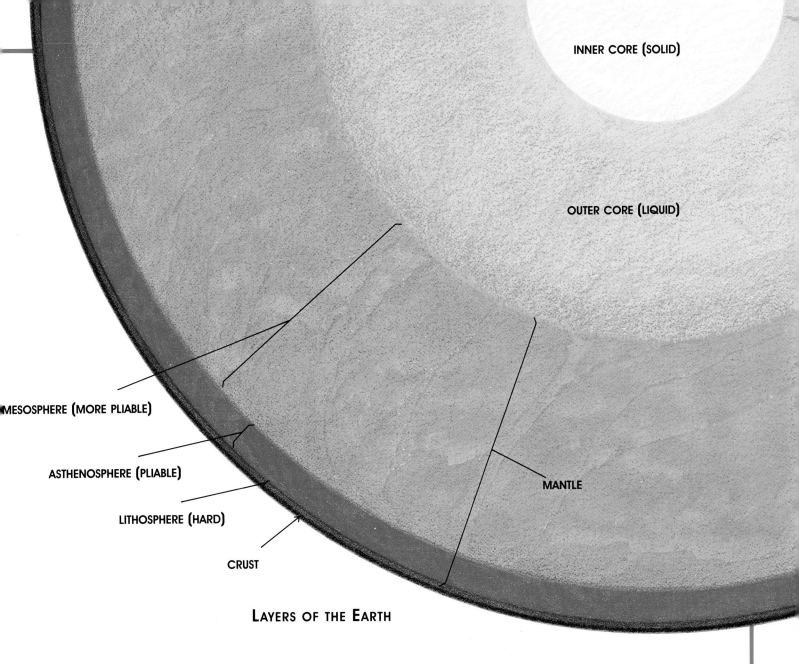

INNER CORE (SOLID)

OUTER CORE (LIQUID)

MESOSPHERE (MORE PLIABLE)

ASTHENOSPHERE (PLIABLE)

LITHOSPHERE (HARD)

CRUST

MANTLE

LAYERS OF THE EARTH

The upper layer of the mantle is colder and harder than the asthenosphere. If it is pulled, it will snap in two with a clean break. The crust and upper mantle together are called the lithosphere, from the Greek word meaning "stony." The lithosphere is the rigid part of Earth. It is 30 to 100 miles (48 to 161 km) thick.

Many clues led scientists to believe the lithosphere is broken up into twenty or more pieces or plates. These plates fit together like the pieces of a patchwork crazy quilt. They float on the asthenosphere like pieces of vanilla wafers on top of chocolate pudding, drifting about the surface of the globe at speeds of 1 to 4 inches, or 2.5 to 10 centimeters (cm), a year. As they do, they pull apart, collide, grind past, or dive underneath one another the way luggage does on a conveyor belt at an airport.

This computer-generated relief view of Earth as seen from over Africa shows both land and undersea topography. To understand plate movement, scientists must study Earth's crust as a single layer, as if it were drained of water, as in the picture on page 41.

Scientists call plate movements plate tectonics, from the Greek word meaning "builder," because it is plate movements that shape the Earth's crust.

The understanding of plate tectonics has helped scientists solve many of Earth's puzzling mysteries. Continents and oceans ride on the plates like passengers on a raft and move about Earth's surface. This explains how the Antarctic region could have had a warm climate at one time. It also explains why evidence of past glaciers is found in the Sahara Desert and why a coral reef lies under the topsoil of my backyard, five hundred miles from the nearest seashore.

TECTONIC PLATES AND THEIR MOVEMENTS

IT IS DIFFICULT TO GET A COMPLETE PICTURE OF EARTH'S continents from its surface. It is even harder to study the 70 percent of Earth that makes up the ocean floor. Some of the best clues to the way plate tectonics works come from pictures taken by satellites stationed high above Earth. Satellite photos give a good overall perspective. Combined with oceanographic information, all of the surface features, both on the land and under the sea, show up very clearly through computer-generated images. The edges of many tectonic plates are also plainly visible.

Photographs taken by satellites can show Earth's surface as well as the edges of its tectonic plates when they are used to create computer-generated images. The boundaries of crustal plates are outlined in yellow in this computer-generated global image.

There are seven major plates on Earth (see pp. 14–15). Each is more than 1,000 miles (1,609 km) across. There may be twice as many smaller ones, some spanning less than 100 miles (161 km). Scientists don't agree on exactly how many small plates exist.

Each plate touches three or more other plates. None can move without affecting all of the others. It is like pushing one domino in a row of upright pieces and making the whole row fall down. The movement of any one plate causes several different things to happen along its edges. Plate movements are responsible for most of the world's volcanoes, earthquakes, high mountain ranges, and deep ocean trenches. All of these are found along plate boundaries or former boundaries.

Not all plates move at the same pace or in the same direction. Scientists learn how fast a plate is moving and the direction it is going by bouncing laser beams off satellites and by recording radio signals from quasars at several widely separated stations.

The plates move much too slowly for us to notice. They have traveled less than 2.5 miles (4 km) in all the time humans have been on Earth. The hour hand of a clock moves ten thousand times faster. Scientists don't know why, but the plates have speeded up or slowed down many times throughout the history of Earth. When the plates speed up, there is usually also an increase in the number of violent volcanoes and earthquakes and in the amount of mountain building at the boundaries.

Fault zones exist where tectonic plates meet. Their names describe the movement of the plates along faults. Here arrows show the directions of stress. Some faults show both vertical (normal or reverse) and sideways (strike slip) movement.

EDGES OF PLATES

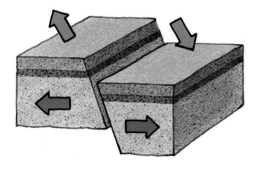

NORMAL FAULT

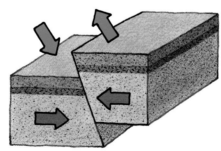

REVERSE OR THRUST FAULT

STRIKE SLIP FAULT

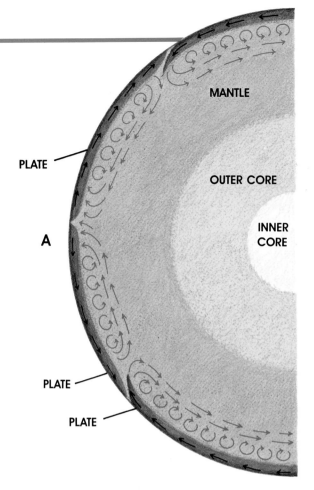

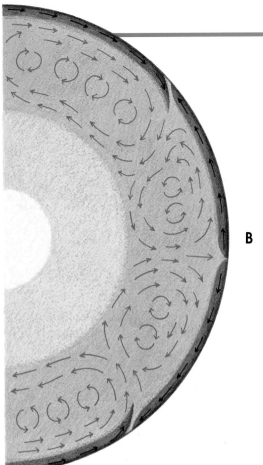

PLATE

MANTLE

OUTER CORE

INNER CORE

A

PLATE

PLATE

B

Figure A: The currents of hot rock that cause plates to move may extend only a few hundred miles into the mantle.

Figure B: The currents may descend all the way to the outer core.

Scientists once thought that hot melted rock swelling up in cracks between plates pushed plates apart and that this was what caused them to move. Most scientists now believe that the plates are moved by convection currents in the asthenosphere. They think the currents circulate through the asthenosphere in much the same way that air circulates in a room. Hot air in a room rises and flows along the ceiling until it cools. Then it sinks to the floor. Currents of hot rock within the mantle may rise until they bump against the bottom of the lithosphere. Some scientists think these convection currents may drag the plates with them as the currents inch along the bottom of the lithosphere.

The currents may be driven by heat flowing from the center of the Earth. Part of the heat comes from decaying radioactive materials. Part is left over from Earth's formation. If the heat were turned off, the plates would stop moving. No new mountains would be built and all old mountains would erode. Earth's surface would look like that of the moon, which has no moving plates: It would be level except for craters made by meteor impacts.

EARTH'S TECTONIC PLATES

Most plates carry both continents and ocean floors. Three, however, carry only ocean floors. The major plates are named for the continent or ocean they carry. The largest is called the Pacific Plate. It covers one-fifth of Earth's surface and stretches from the western coast of North America to the Philippine Islands. It carries only the Pacific Ocean floor, its islands, and a narrow coastal strip of California on the eastern side of the plate. It moves toward the northwest in a roughly counterclockwise rotation at about $2^1/_2$ inches (6 cm) a year—about as fast as your fingernails grow.

The Pacific Plate touches seven other plates. A lot of action takes place along its boundaries. It slides past the American Plate, pulls away from the Antarctic, Cocos, and Nazca plates, and dives under the Indo-Australian, Eurasian, and Philippine plates.

The least amount of activity takes place where the Pacific Plate slides past the American Plate. Scientists call this a transform boundary. About the only things that happen at transform boundaries are earthquakes, and earthquakes happen at all plate boundaries. Any movement of the plates can trigger a quake, because plate movements place great stress on rocks within the lithosphere.

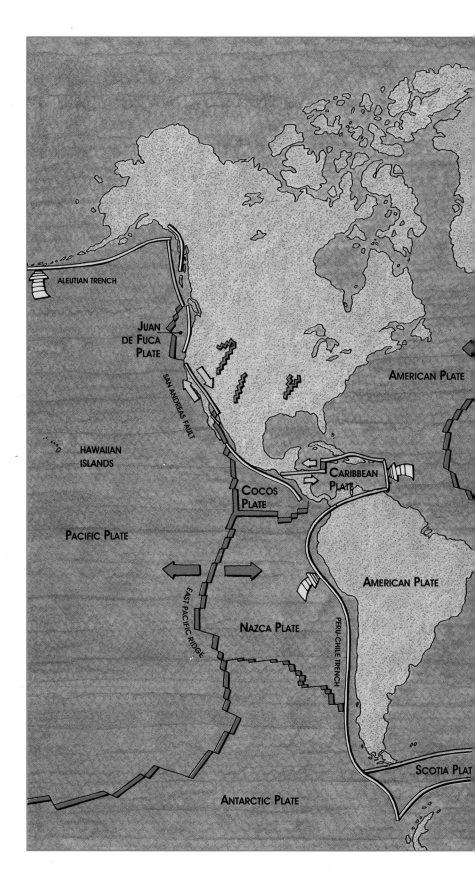

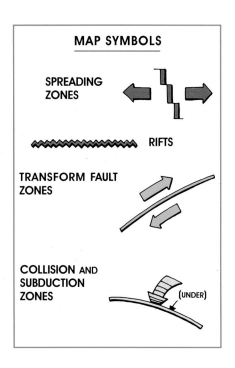

MAP SYMBOLS

SPREADING ZONES

RIFTS

TRANSFORM FAULT ZONES

COLLISION AND SUBDUCTION ZONES (UNDER)

When lithospheric rock comes under too much pressure or stress, it fractures suddenly, creating a fault. In an ordinary quake, the break happens very quickly. A fault can spread more than five-eighths of a mile (1 km) in a second.

If the edges of tectonic plates were smooth and straight, there would be much less action at transform boundaries. Unfortunately, this is seldom the case. The rock along the edges of plates has many bulges and bends in it. As one plate slides past the other, it gets hung up at one of these rough spots. This causes great strain and stress to build up, and eventually the rock breaks or fractures. The resulting fault creates an earthquake, which often is damaging. It can destroy buildings 10 to 100 miles (16 to 161 km) away.

The break causes Earth to tremble and sends seismic waves rippling in all directions. Like the ripples on a pond when a rock is tossed into it, seismic waves travel outward from the epicenter. The violent shaking usually lasts a minute or less. The longest known earthquake lasted only seven minutes, and the 1989 Loma Prieta quake, in California, lasted just six seconds.

Scientists measure the strength of a quake with a magnitude scale. Every earthquake is given a magnitude to indicate how violent it is. Just as the colder temperatures are given minus numbers (below zero), the smallest detectable earthquakes have magnitudes of less than zero. On the most commonly used scale (the body-wave magnitude scale), an earthquake of magnitude 2 is thirty times larger than one of magnitude 1. A magnitude 3 earthquake is thirty times larger than a magnitude 2 quake and nine hundred times larger than a magnitude 1 earthquake.

The Richter scale, which is most often referred to in news articles in the United States, is a magnitude scale designed for a particular kind of seismograph called the Anderson seismograph. This seismograph is seldom used anywhere except in southern California; scientists do not use the Richter scale to record the magnitude of earthquakes anyplace where a different type of seismograph is used. All magnitudes given in this book are universal body-wave magnitudes.

The smallest earthquakes that can be felt by people are magnitude 2.5 to 3.0. The largest have exceeded magnitude 9. Earthquakes larger than magnitude 10 have seldom, if ever, occurred in recorded history.

This actual seismographic reading shows vertical ground motion over 11 minutes just after the January 1994 Northridge, California, earthquake. The 6.8 Richter magnitude quake was detected by a seismometer at the bottom of a 394-foot-deep (120 m) borehole and recorded at the Oklahoma Geological Survey Observatory near Leonard, Oklahoma, 1,291 miles (2,078 km) from the epicenter. The first arrival, about 240 seconds after the quake, is a **p**ush-pull **P** wave traveling through Earth's upper mantle. **S** is a wave with shake motion, also in the upper mantle. **LR**, a **R**aleigh surface wave that spreads along the Earth's surface like ripples in water, is strongest 720 seconds after the quake. In the first two months following the quake itself, more than 600 aftershocks were documented; many others occurred that were not monitored.

1994 JAN 17 NORTHRIDGE CA Ms=6.8

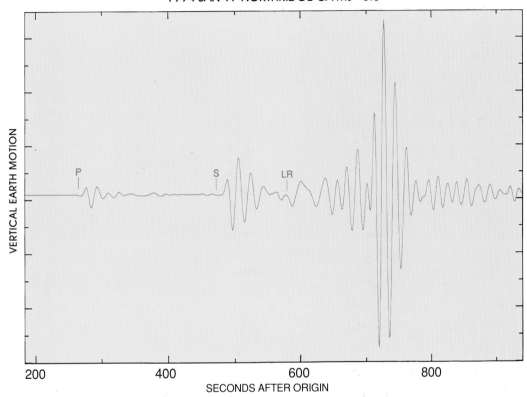

SECONDS AFTER ORIGIN

The magnitude of a quake depends on how large the ruptured fault was and how great the strain was before the break. Usually, the greater the strain, the longer the rupture and the bigger the quake. After the initial break, the tension is released. The fault is quiet and inactive for a while, until stress builds up again, although aftershocks may occur due to slippage along the fault after the rock breaks.

The transform boundary between the Pacific Plate and the American Plate is marked by the San Andreas Fault. It runs across central California's Carrizo Plain and can be clearly seen from the air. The land on the western side of the fault is a passenger on the Pacific Plate. The land on the eastern side is carried by the American Plate.

The San Andreas Fault stretches 650 miles (1,040 km), from the southern tip of the Gulf of California to just north of San Francisco. In 1906 an enormous break along this fault caused a magnitude 8.6 earthquake that devastated San Francisco (*left*). Sand and rubble from the destroyed buildings were used to fill a lagoon that became known as the Marina District. Ironically, this landfill liquefied during the 1989 Loma Prieta quake, increasing damage to the Marina District (*right*) and spewing up sand, burned wood, and other rubble from 1906 that had been buried for three generations.

Millions of years of butting and bruising have shattered California on both sides of the San Andreas Fault with hundreds of fractures and faults that have thrust up mountains and interrupted the flow of streams. In 1971, a rupture along the San Fernando Fault thrust parts of the mountains along its edge 8 feet (2.4 m) higher.

There is another transform boundary on the opposite side of the Pacific Plate. This one splits New Zealand. The east side of the island is carried by the Pacific Plate, which is sliding southwest. The west side is carried by the Indo-Australian Plate, which is sliding northeast. Slippage along this boundary caused a damaging earthquake in New Zealand in 1989.

The Dead Sea Fault zone is also a transform boundary. In this area the African Plate is moving past the Arabian Plate. Farther south, these two plates are pulling apart.

RIFTING AND DRIFTING

FROM THE TOP OF A STEEP CLIFF I GAZED ACROSS THE GREAT RIFT Valley to an equally steep cliff on the other side. It was an exciting moment for me. Although I couldn't actually see any movement, I knew that at that place and at that moment, the movements of two tectonic plates were ripping the crust of east Africa apart like the seam of an old shirt. Just as Africa was once pulled away from South America, the Arabian and Somali plates are now slowly pulling away from the African Plate. It started millions of years ago when the crust stretched too thin and two great cracks in Earth's crust began to open up, causing the floor of the Great Rift Valley to slowly drop hundreds of feet. The earthquakes that were generated by these enormous dip slip faults must have been tremendous.

Of course, there is no witness for a process that lasted hundreds of years, but there is plenty of evidence to show what happened. The rift valley system stretches from the northern end of the Red Sea to Mozambique. The cliffs on both sides are lined by volcanoes, some long dead, some still active. Scientists have positive evidence that the floor dropped. In some places, strata (layers) of rock on the valley floor exactly match strata in the rock at the top of the rift walls directly above.

Scientists call the places where two plates pull apart "spreading boundaries." Spreading boundaries are easy to find. They show up plainly on satellite pictures. The action that takes place there includes eruption of volcanoes, building of mountain ridges, and earthquakes.

Spreading boundaries occur in narrow belts rimming the edges of both plates. Most of them are found on ocean floors because new oceans are opened up when continents tear apart. The Red Sea is a future ocean just beginning to be formed where the Arabian Plate is pulling away from the African Plate.

300 MILLION YEARS AGO   200 MILLION YEARS AGO   100 MILLION YEARS AGO   50 MILLION YEARS AGO

The forces tearing Africa apart are the same ones that have been breaking up continents for at least a billion years. Scientists believe that all of Earth's lands have been joined together into one huge supercontinent many times throughout the history of the planet. The last supercontinent, which scientists call Pangaea, began breaking up about 200 million years ago. Spreading boundaries created Laurasia and Gondwanaland, then broke those two continents apart as well.

Scientists think that spreading boundaries may be caused by the convection currents of hot magma rising under the lithosphere plate. This magma comes from deep within Earth. When it rises, it presses against the lithosphere, stretching it and making it become very thin. Finally, as the two plates continue to slowly pull apart, it cracks. Orange-hot magma rises up to fill the gap much the same way that the white of a cracked egg oozes out when it is boiled. Volcanoes may erupt, creating new crust and a ridge of mountains along the seam, or rift.

Scientists can discover how long plates have been pulling apart by studying the lava on either side of the crack. The youngest lava lies nearest the rift; the oldest is farthest.

This cutaway view of Earth shows hot magma oozing up through a midocean rift.

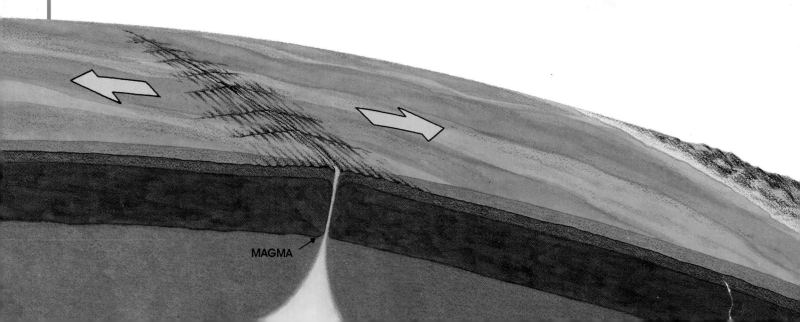

MAGMA

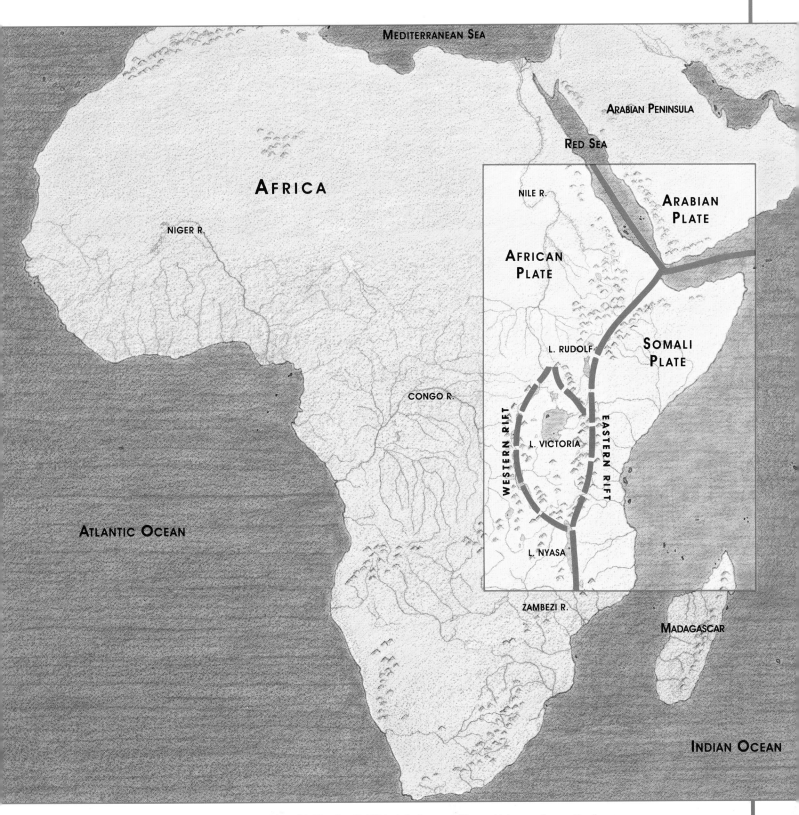

MEDITERRANEAN SEA

ARABIAN PENINSULA

RED SEA

AFRICA

ARABIAN PLATE

NILE R.

AFRICAN PLATE

NIGER R.

SOMALI PLATE

L. RUDOLF

CONGO R.

WESTERN RIFT

L. VICTORIA

EASTERN RIFT

ATLANTIC OCEAN

L. NYASA

ZAMBEZI R.

MADAGASCAR

INDIAN OCEAN

East Africa's Rift Valley stretches 3,500 miles (5,600 km). Its furrows, cliffs, and lakes are formed by three plates' slowly pulling away from one another.

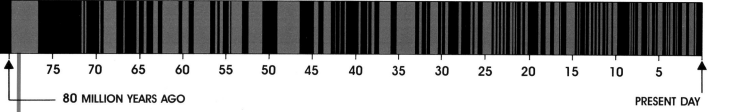

75 70 65 60 55 50 45 40 35 30 25 20 15 10 5

80 MILLION YEARS AGO   **PRESENT DAY**

Rock that has been melted can be dated by its magnetic alignments, or directions, because the Earth is like a giant magnet with north and south poles. Magma and lava contain small grains of minerals that act like needles on a compass. One end always points toward Earth's magnetic North Pole and the other toward the South Pole. These grains are frozen in place when the molten rock cools and hardens. They are called paleomagnetic markers.

For some unknown reason, the North and South poles have switched their magnetic polarity, flipping back and forth many times in the history of Earth. Three hundred times in the past 200 million years, north became south and south became north.

As a result, zebra-striped bands of lava rock whose magnetic grains point in opposite directions cover the entire ocean floor on either side of a rift. They record magnetic reversals something like the way an answering machine records messages on a tape. These bands are easily detected by magnetometers towed behind aircraft and ships.

The two scales show how Earth's magnetic fields have reversed themselves during the last 80 million years. The fields are generated by the motion of molten iron and other elements in the outer core.

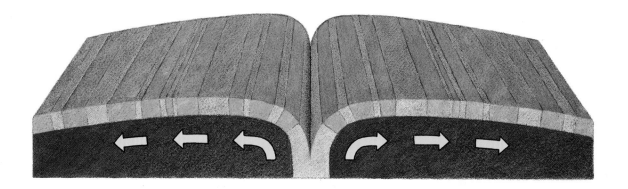

The Atlantic Ocean was formed when the American and Eurasian plates pulled apart. Both plates carry continents as well as oceans. The American Plate stretches from the San Andreas Fault to the middle of the Atlantic Ocean. It is moving west-southwest about one-half inch, or 12 millimeters (mm), a year. The Eurasian Plate is moving toward the east-southeast at about the same speed. It includes the eastern half of the Atlantic Ocean and the continents of Europe and Asia. In your lifetime these two plates will have separated by about your adult height. European settlers have to travel about 10 yards (9 m) farther today to reach the shores of North America than they did in the 1600s.

Some plates are moving apart, some are being forced together, and some are sliding past one another. Study of the magnetic stripes on the Atlantic floor shows that the American and Eurasian plates are moving away from each other at a little less than 1 inch (2.5 cm) a year. The Pacific Plate and Nazca Plate are separating at about 7 inches (17.5 cm) per year—eight times faster than the American and Eurasian plates are pulling apart. *(Map shows relative movement in cm per year.)*

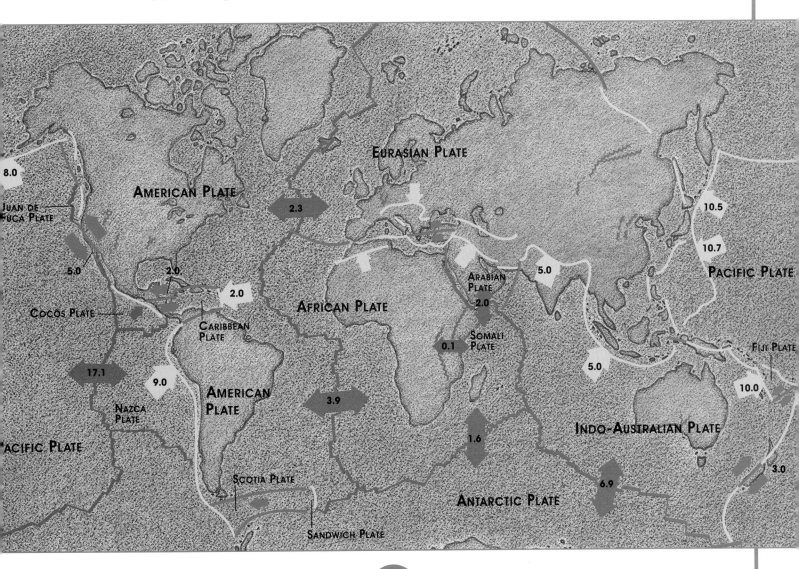

The ridge of volcanic mountains along the rift between the American and Eurasian plates is known as the Mid-Atlantic Ridge. It is the longest and tallest mountain range in the world. This 12,000-mile-long (20,000 km) ridge is a part of a 46,000-mile (74,000 km) underwater mountain chain, formed by spreading boundaries, that winds around the globe like the seam of a baseball. Some of its peaks are 36,000 feet (11,000 m) high.

Iceland provides the best visual evidence of activity at spreading zones. This volcanic island straddles the Mid-Atlantic Ridge and is a part of it. It grows wider every year as the two plates it rides on pull apart. Icelanders experience almost continuous volcanic action and are frequently shaken by earthquakes as stress from the separating plates fractures rock within the lithosphere. ▶

The island of Surtsey, south of Iceland, was born November 14, 1964, when undersea volcanic activity reached the ocean's surface. The eruption, here sixteen days old, continued for three and a half years. ▼

Not all hot spots become rifts. Sometimes a single column of hot melted rock burns through the lithosphere in the middle of a plate. It does not pull the plate apart. It only builds a volcano. Plumes of molten rock such as these are stationary; they do not move. However, tectonic plates drift over the hot spots, and so the plumes form chains of volcanic islands such as the Hawaiian Islands. It is possible to follow the direction in which the Pacific Plate has moved by following the trail of volcanic islands left by the plate as it drifted over the Hawaiian hot spot.

Yellowstone National Park, which sits over a pool of hot magma that lies just below the surface, may be the site of a future separating point. This spot is believed to be the huge crater, or caldera, of an ancient volcano. It is part of a 1,200-mile-long (1,931 km) rift in the western United States called the Great Basin. Heat from deep within the Earth began thinning and stretching the crust in this region 10 to 15 million years ago. At that time the areas now occupied by Salt Lake City and Reno were about 250 miles (402 km) closer together than they are today.

Every ten thousand years or more, a large block of land sinks as the rift continues to pull apart, creating long valleys (such as Death Valley) between steep mountains. A woman hunting elk in Idaho in 1983 witnessed one of these events. She saw and heard the mountain across from her crack along a fault and drop nearly 10 feet (3 m). "It sounded," she said, "like a sonic boom." The dropping of the mountain caused a magnitude 7.3 earthquake.

Another active rift, the Rio Grande Rift, stretches from Colorado to Texas and is marked by the Rio Grande River. This six-mile-deep (9.7 km) crack in the planet is hidden by deep sediments and covered by a layer of lava. Santa Fe and Albuquerque sit atop this rift, which is growing a bit larger every year. Someday these cities may be victims of earthquakes or volcanoes.

Geologists are not sure whether either the Great Basin area or the Rio Grande Rift will someday separate enough to open a new ocean. If they do, the West Coast will become a free-floating continent.

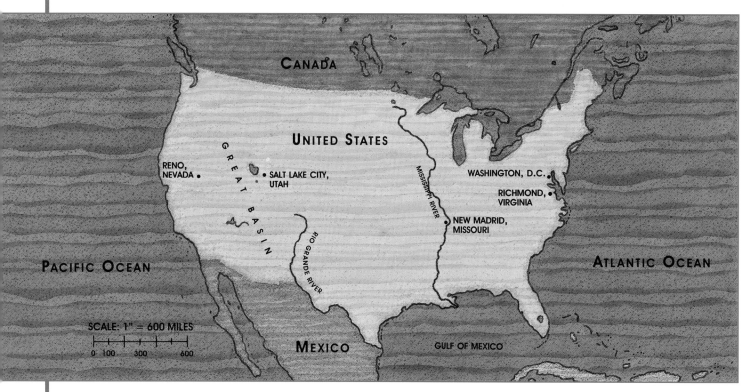

In 1811 and 1812 three earthquakes with estimated magnitudes of 7.2, 7.1, and 7.4 occurred along the zigzag New Madrid Fault. One segment of the fault broke during each quake. It is said that the ground moved in oceanlike waves and opened cracks too wide and deep for horses to cross. The quakes were felt as far north as Canada. They broke plaster in Richmond, Virginia, and rang bells in Washington, D.C.

Scientists think that the Mississippi River Valley was formed by a rift that failed to completely separate about 600 million years ago. It was inactive for a long time, but stress from plate activity has reactivated part of it. The area around New Madrid, Missouri, is now one of the most active earthquake regions east of the Rocky Mountains. It averages one small quake every 48 hours. It is also the site of three of the greatest earthquakes in eastern North America.

The faults that cause the New Madrid quakes are deeply buried under river sediments and a thick layer of marine sediments that were deposited when the area dropped below sea level.

Something stopped this rift before it could form a new ocean. The rift forming the Mid-Atlantic Ridge, on the other hand, is still very active, and the Atlantic Ocean grows wider each year. As it does so, the Pacific Ocean grows smaller. Another kind of plate action is created as the western edge of the Pacific Plate collides with the Indo-Australian, Eurasian, and Philippine Plates.

COLLISION BOUNDARIES AND CONTINENT BUILDING

ON THE MORNING OF MAY 18, 1980, MOUNT SAINT HELENS erupted with an explosion five hundred times greater than that of the atomic bomb dropped on Hiroshima. This Washington State volcano is a direct result of a collision between the Juan de Fuca Plate and the American Plate.

Scientists call the places where two plates collide convergent, or collision, boundaries. More activity takes place at collision boundaries than at other boundaries and the actions are more violent. Eighty percent of the world's volcanoes, 90 percent of all earthquakes, and most of the world's mountain building occur along collision boundaries or along the edges of formerly colliding plates.

This eruption of New Zealand's most active volcano, Ngauruhoe, occured in January 1974. A cloud of volcanic fragments can reach temperatures of 1000 degrees Centigrade and speeds of 93 miles (150 km) per hour.

When two plates collide or converge, one is shoved under the other. Deep trenches develop as the lower plate bends downward and dives underneath the upper one. These places are called subduction zones.

Scientists suggest that subduction zones form where convection currents cool and sink. They think the lower plate is dragged down, or subducted, into the asthenosphere, where it melts and is recycled.

The deepest trench on Earth is located where the Pacific Plate dives beneath the Philippine Plate. It is 6.5 miles (10.8 km) deep, more than six times as deep as the Grand Canyon.

The Farallon Plate disappeared beneath the American Plate 85 million years ago. The Rocky Mountains, which were pushed up by the collision, are the only visible evidence of Farallon's existence. Some scientists think that the melting of this subducting plate may be the source of heat that is still stretching and thinning the Great Basin today.

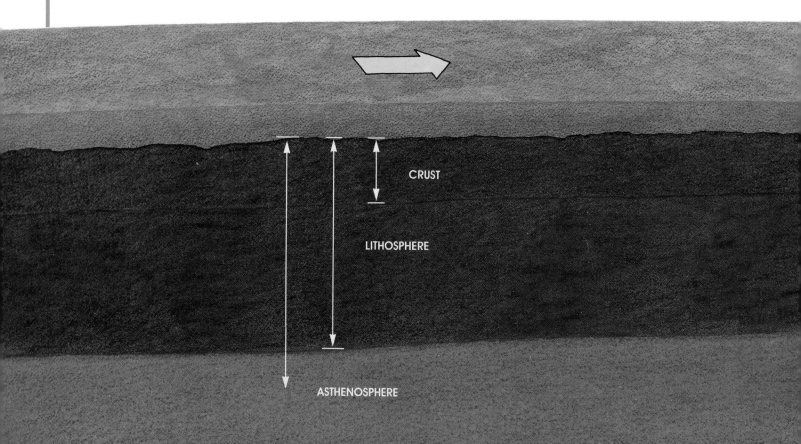

CRUST

LITHOSPHERE

ASTHENOSPHERE

In collisions between plates carrying oceans and those carrying continents, the subducting plate is always the one carrying the ocean, because ocean floors are heavier than continents. Mountains formed by such collisions are called uplift mountains. Sometimes the melted rock breaks through the surface and erupts as volcanoes — as it did at Mount Saint Helens — along the edge of the overriding plate. These volcanoes tend to be more explosive than those along separation boundaries.

Most of the melted rock, however, wells up under the edge of the overriding plate. It cools and hardens below the surface after raising great mountain ranges or plateaus, such as the Rocky Mountains. Several plates are responsible for lifting the Andes Mountains in South America. The Nazca Plate continues to elevate the Andes one inch (2.5 cm) every year.

The Pacific plate shrinks every year because it is subducting beneath the Indo-Australian, Eurasian, and Philippine plates. The Nazca Plate is becoming smaller as it dives beneath South America. This plate may eventually vanish, because it is subducting faster than it is growing at its spreading points. This picture shows how one plate is forced under the other at a subduction zone.

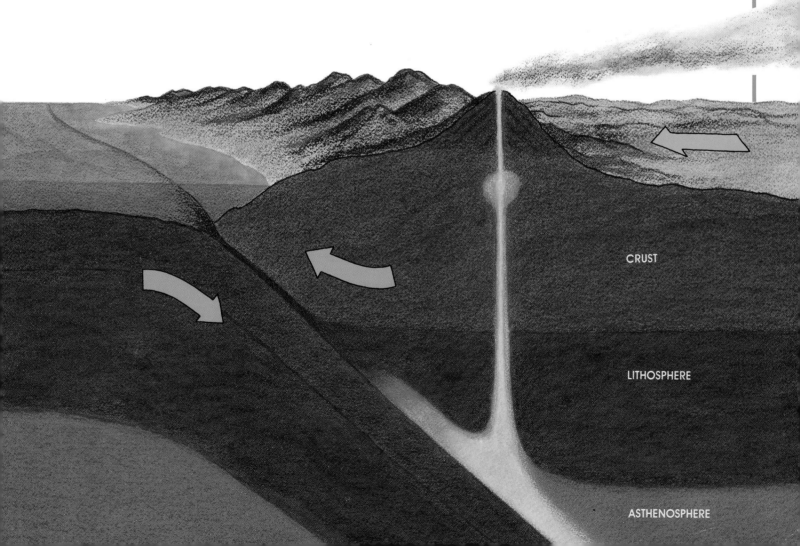

CRUST

LITHOSPHERE

ASTHENOSPHERE

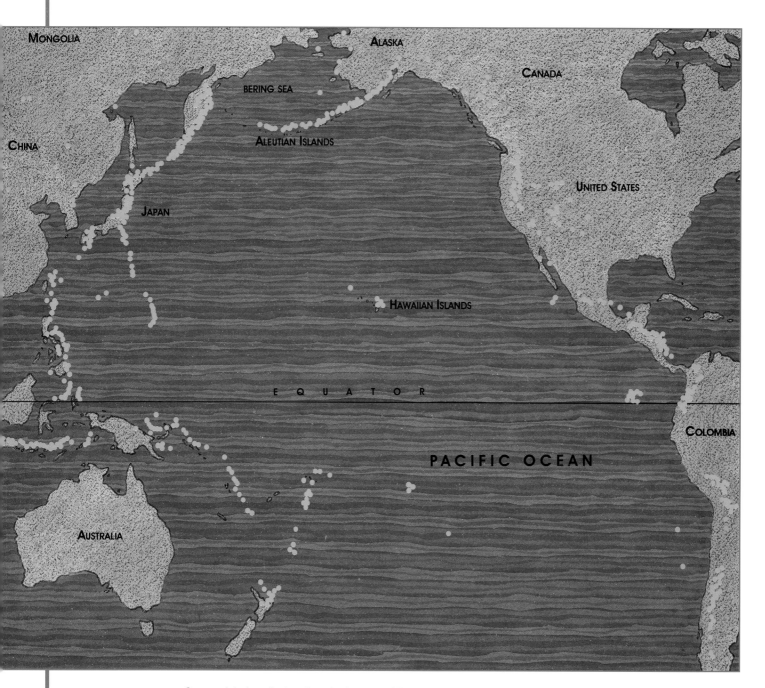

Orange dots show the location of volcanoes at the edge of the Pacific Ring of Fire. They are situated at subduction zones that circle the Pacific Ocean.

If both of the colliding plates carry only ocean, either one may subduct. Volcanoes erupt along the edge of the overriding plate. Their cones make a chain of islands, or an island arc. The Aleutian Islands of Alaska form an arc where the Pacific Plate dives beneath the American Plate along the edge of the Bering Sea.

Scientists think that our present continents may have been built from island arcs that were swept together at subduction zones, like floating toys in a bathtub after the drain plug has been pulled. Since islands also are too light to subduct, they collided and stuck, freezing together like ice cubes in a pitcher of Kool-Aid. Eventually they grew into minicontinents.

Scientists call these minicontinents cratons. Most of them are more than 2.5 billion years old and are the oldest and strongest parts of today's continents. Some are half as big as the United States, others are smaller than Texas. Plate movements welded the cratons together into cores of huge continents.

Continents may look as if they are single pieces of solid land when one drives across or flies over them. Actually, every continent is more like a collage with many pieces stuck together. Two to three billion years ago there were only islands, underwater mountain ranges, and a few minicontinents on Earth. Collisions between tectonic plates built big continents out of these bits and pieces.

It is believed that the core of North America was formed about 1.8 billion years ago when seven cratons joined together. Scientists have found evidence of one such union. Ancient volcanoes and the remains of a high mountain range, buried two miles (3.2 km) below the surface, stretch across Ohio, Indiana, Illinois, and Missouri. Scientists believe these mountains were created by a collision between two cratons 1.4 billion years ago.

Every island, undersea mountain or plateau, and piece of ancient continent carried by subducting oceanic plates eventually gets scraped off and plastered to the leading edge of a continental plate, greatly increasing the size of the continents. Scientists call this plastered-on material terranes.

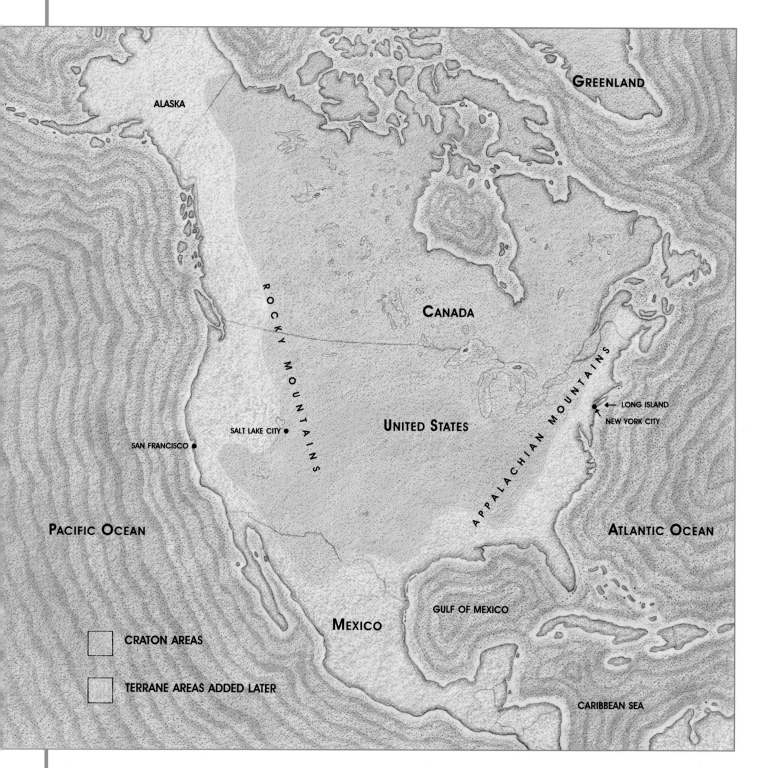

ALASKA

GREENLAND

CANADA

ROCKY MOUNTAINS

UNITED STATES

APPALACHIAN MOUNTAINS

SALT LAKE CITY ●

SAN FRANCISCO ●

← LONG ISLAND
↑
NEW YORK CITY

PACIFIC OCEAN

ATLANTIC OCEAN

MEXICO

GULF OF MEXICO

CARIBBEAN SEA

☐ CRATON AREAS

☐ TERRANE AREAS ADDED LATER

Two hundred fifty million years ago, present-day Salt Lake City, Utah, would have been near the western edge of North America. Most of the land west of there and east of the Appalachian Mountains is made up of hundreds of plastered-on terranes. Fossil remains and ancient magnetic traces show that some of these terranes came from as far away as the equator.

Sometimes strain placed on plates weakens rock and causes the joints between terranes to break, creating faults and earthquakes in the middle of a plate. New York City may be sitting on such a fault, or sheer zone, where land masses are sliding past each other at about half an inch (1.27 cm) a year. If it continues for a million years, Long Island may be split in two. Scientists think this may be a spot where a plate boundary used to be. A small quake occurred there in 1987. Scientists predict that someday there will be a strong quake in that area.

In a lettuce field near El Centro, California, the displacement of the rows as a result of a magnitude 6.9 earthquake is a dramatic illustration of a fault.

Plate movements didn't stop with creating large continents. Many times throughout the history of Earth, they brought all of the continents together, forming huge supercontinents. Then they broke them up to be reformed again later on. In the process, they left bits and pieces behind. Florida, for example, was once a piece of Africa, and Scotland was once a part of North America.

When two continents collide, their leading edges crumple like two cars in a head-on collision. They form folded mountains — long parallel ranges like those of the Appalachian Mountains of North America and the Ural Mountains between Europe and Asia. The Urals mark the place where two huge land-bearing plates collided in the distant past and formed Eurasia. The Appalachian and Ouachita mountains in Arkansas and Oklahoma are scars left when Eurasia and Africa slammed into North America millions of years ago. Before erosion wore them down, these mountains were as high as the Alps and Himalayas. The Alps were thrust up when the African Plate rammed into Eurasia.

The towering peaks of the Himalayas in Nepal are shown above left. Tenache, in the foreground, is 13,000 feet (3,961 m) high; in the background is Ana Dablang, 22,500 feet (6,857 m) high. They seem dwarfed when viewed from a satellite perspective, such as the 1988 Soviet *Aragatz 2* mission photograph shown below.

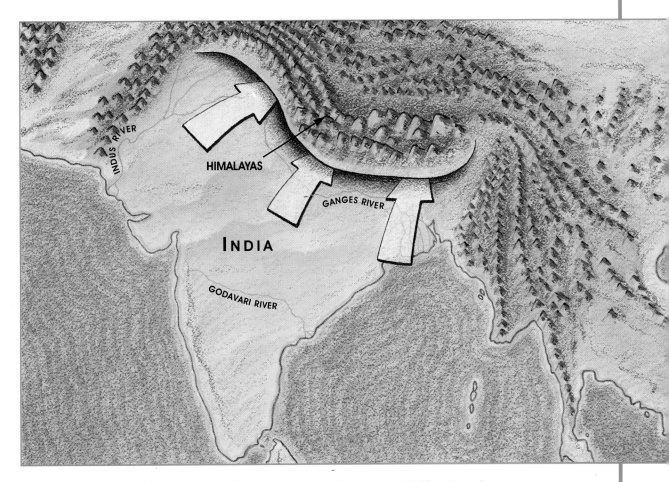

The Himalayas are a result of the Indo-Australian Plate grinding into Asia. This collision buckled the edges of both plates and folded them into huge crustal waves. Seashells from along the coasts of the plates were lifted 2,000 feet (600 m) above sea level. These 4-mile-high (6.4 km) mountains, the highest ones on land, are still growing 2 inches (5 cm) every year.

As the Indo-Australian Plate collides with the Eurasian Plate, India continues to shove into and under Asia and many devastating earthquakes shake the ground far inland. Tectonic plates are cold and brittle, like china dinner plates. As they subduct, the pressure to bend becomes too great and they break, causing the ground above to shake. Many of these earthquakes have magnitudes as high as 8 and cause great damage. The plates continue to break periodically until they plunge deep enough and grow hot enough to flow. Then the quakes stop, but this may take ten million years.

In 1985, the subduction of the Cocos Plate beneath the American Plate caused a magnitude 8.1 earthquake that destroyed much of Mexico City when the plate snapped 12 miles (19 km) below the surface.

Along the western coast of Central America, where one tectonic plate dips beneath another, more than 35 earthquakes of magnitudes greater than 7.0 have occurred in this century, one of the most recent in May 1994. The quake of 1985 released an estimated 1,000 times as much energy as the Hiroshima atomic bomb. Nevertheless, many buildings in Mexico City were not built to resist earthquakes. This picture shows one of the 800 buildings that collapsed, while a 44-story structure and a transmission tower, designed to withstand seismic activity, still stand in the background.

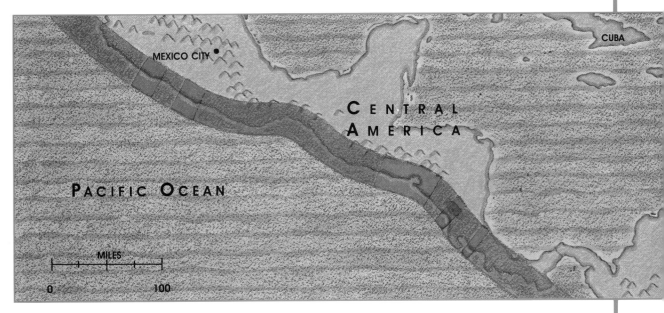

Blue shading shows where tremors have relieved pressure along fault lines within the last forty years. Red shading shows where pressure is still building without relief.

Some scientists think that the northwest coast of North America may be storing energy for a similar quake. The Gorda, Juan de Fuca, and Explorer plates are subducting under the American plate in this area, but there hasn't been a large quake in the region for more than 200 years. One is long overdue.

More gigantic, damaging earthquakes occur in Japan, China, and Chile than anywhere else on Earth. Japan sits directly over the place where the Pacific Plate is burrowing under the Eurasian Plate. The small Nazca Plate is responsible for the large number of disastrous earthquakes in Chile.

One of the largest quakes in recent years caused no damage at all because the epicenter of this magnitude 8.3 earthquake was on the bottom of the Pacific Ocean. It struck south of New Zealand along the boundary between the Pacific and the Indo-Australian plates. Scientists think that this section of the boundary may be changing from a transform boundary to a subduction zone. They are very excited about it, because if it is, by studying that region they may be able to learn how subduction zones are formed.

Scientists can't prevent earthquakes or volcanoes. But they are studying ways to predict them and to defuse them so that they won't cause quite so much damage.

PLATE MOVEMENTS AND OUR FUTURE

SCIENTISTS HAVE LEARNED MUCH ABOUT TECTONIC PLATES AND how they affect our lives. We know that they cause earthquakes and volcanoes. We also know that where these disasters have occurred, they will occur again. They are so frequent around the rim of the Pacific Ocean, where several plates interact, that scientists call it the Ring of Fire. But knowing where they will occur and why isn't the same as knowing exactly when they will happen.

Because it is a zone of frequent earthquake activity, the edge of the Pacific Plate is known as the Pacific Ring of Fire. ▲

◄ Although they are destructive, volcanic eruptions are also beneficial: They provide sources of rich, fertile soil, construction material, and inexpensive power. From the study of lava flows such as this one in Hawaii, geologists can learn more about their impact on the environment.

Seasat is equipped especially to record and transmit information about the oceans from high above the surface of Earth.

Alvin is only one of the submarine laboratories in which scientists can gather information directly from the ocean floor.

Many scientists are trying to solve this mystery. Seismologists feed information about earthquakes into computers to make detailed, X-ray-type images of Earth's interior. Geophysicists study the effects of heat and pressure on rocks. They lower themselves into volcano craters and walk across lava fields to find out how melted rock behaves.

Geologists study rock formations to learn what happened on Earth in the past. In areas where much of the rock is bare of plants and top-soil, they can clearly see where rocks violently crashed together billions of years ago.

Oceanographers dive to the bottom of the ocean in tiny submarines to see what is going on at separation zones and midocean ridges. They also map the ocean floor, taking sonar pictures by bouncing sound waves off underwater objects.

Although scientists understand plate movements and know how earthquakes and volcanoes work, they still cannot predict exactly when an earthquake or volcanic eruption will take place. But they can prevent some of the damage these natural disasters cause. Scientists have discovered some clues that suggest that an earthquake is about to happen.

There are events that have occurred over and over again just before earthquakes struck in the past. For example, in areas where stresses were building, there were many more microearthquakes (quakes that show on a seismograph but cannot be felt) than usual. Animals in the region often behaved strangely, perhaps because they sensed the microearthquakes. Also, the amount of radon in well water built up for weeks or months and then declined almost back to normal just before the earthquake occurred. The level of water in wells seemed to drop, then rise. An increase of hydrogen in the soil was noticed, and sometimes a gentle tilting of land near a fault zone was detected.

Using computers and gravitational imaging information from satellites, scientists draw detailed maps of the Earth. They use computers to construct models of how the continents were assembled, torn apart, and reassembled; and they calculate what might have caused these events. They also attempt to predict future occurrences. This model shows Earth's surface in black and white; plate boundaries are yellow. Red indicates earthquake epicenters for magnitudes greater than 5 for the decade 1980–1990. Such maps can be used as a tool for predicting future seismic activity.

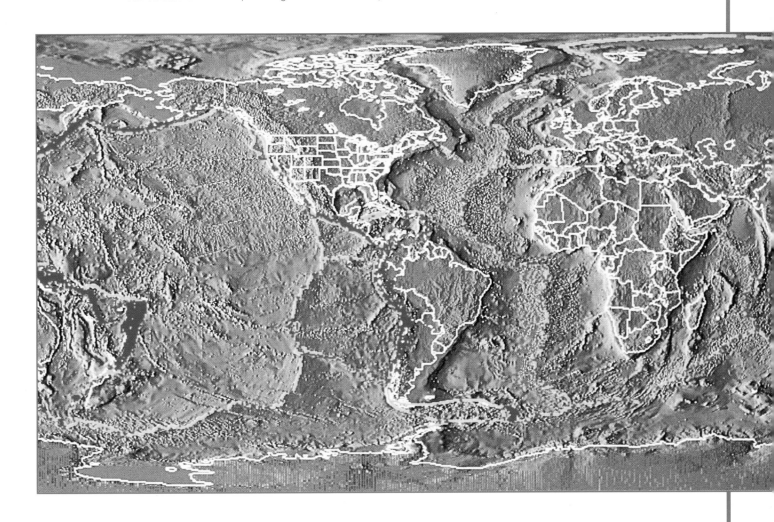

Scientists who study the Earth's magnetic field have discovered something else that might prove to be useful in predicting earthquakes. They have detected increases in electromagnetic waves during the weeks and hours before some earthquakes struck. It is an exciting discovery, but only time and further study will determine whether this increase always happens before earthquakes occur.

In the meantime, scientists will continue to use proven clues to save lives. In 1975, just after a swarm of small tremors hit a region in China, scientists noticed a variation in the amount of radon in the well water. They also received a large number of reports of unusual and erratic behavior of animals. Just to be safe, officials ordered everyone in the area to leave their homes. Although a strong earthquake destroyed most of the town's buildings the next evening, very few lives were lost.

Most people who die in earthquakes are killed by falling buildings or other man-made structures, not by the quake itself. Earthquakes themselves do not usually cause death. This is a message that scientists want to get across to everyone who lives in an earthquake zone. They want to teach these people to construct earthquake-safe buildings. Such buildings don't cost much more to build than unsafe ones.

In 1970, 250,000 people were killed during an earthquake in China. In a similar quake in Chile, only 150 people died. The buildings in Chile were strong enough to withstand the quake; those in China were not. Only sixty buildings were destroyed in San Francisco during the 1989 earthquake, mainly because the city has very strict building codes requiring that all new buildings be designed for earthquake safety. More lives were lost in the United States due to the effects of severe winter weather in 1993–94 than because of the Northridge, California, earthquake.

Scientists are also planning to teach the people in both volcano and earthquake zones to set up warning systems similar to those being developed in California along the San Andreas Fault and in Washington. Advance warnings and evacuation saved the lives of thousands of people when Mount Saint Helens erupted. Fewer than 100 died. On the other hand, a much less powerful volcano in Colombia killed 25,000 people because they had no warning and were not evacuated.

Active mass dampers are already used in some skyscrapers to neutralize the effects of high winds or earthquakes. A mass damper is a dense, heavy weight that is suspended from cables or mounted on tracks near the top of the building. When the building sways, a computer causes the weight to be shifted in the opposite direction so that the structure remains relatively steady.

Although the Chinese and Colombian events could not have been predicted, they should have been expected — and planned for — because both areas lie within the Ring of Fire. We know for sure that eruptions and quakes will happen there again. If people live in danger zones, they must learn to live with the existing natural forces. Many deaths can be prevented in the future if the people are properly prepared.

Many of the large cities on Earth are located in earthquake zones. Sadly, most of these cities do not have earthquake-proof buildings. In fact, few of the buildings in Los Angeles, California, or Memphis, Tennessee, could withstand a strong earthquake even though it is known that both are located near active fault zones and are almost sure to have a major quake sometime in the future. Earthquake-safe buildings in these cities could save the lives of thousands.

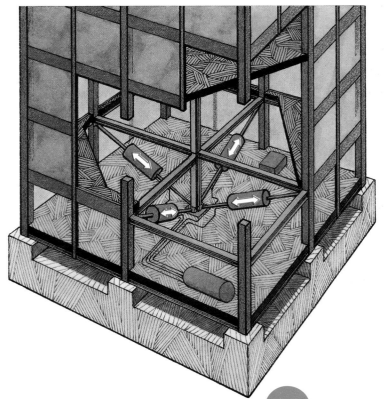

Active tendon systems work on the same principle as active mass dampers. Instead of moving a mass at the top, though, a computer moves the building with pistons at the foundation.

Tectonic movements bring destruction, but not everything about them is destructive. Their shifting and jostling are responsible for many of the valuable mineral deposits on Earth. Most of those found on land are located near the edges of continents where plates have subducted or collided.

The deposits found during the Alaskan gold rush were formed when an island arc smashed into the continent near Seward. Most of our copper, nickel, barite, and some lead, iron, and silver deposits are also products of former volcanic islands that were plastered to the continents. Diamonds, on the other hand, were brought to the surface by volcanoes. They were formed at the bases of continental plates.

Plate tectonics also created some of our gas and oil deposits. Scientists believe that the heat and pressure of ancient subduction zones killed the plants and animals that were living along the seashores. After millions of years of stewing, the oil and gas became trapped in pools in porous rocks.

Scientists are using their knowledge of plate tectonics to search for ancient subduction zones. They hope to find rich oil fields and other mineral deposits in them.

This picture shows the likely location of minerals produced by the movements of tectonic plates. Most of the minerals found on land are near the edges of continents where plates have subducted or collided. In addition, downward warping at subduction zones produces large, deep basins. Marine life-forms and sand washed down from nearby areas above the sea fill these basins with sedimentary rocks, and eventually heat and pressure convert dead marine creatures and plants into oil and gas.

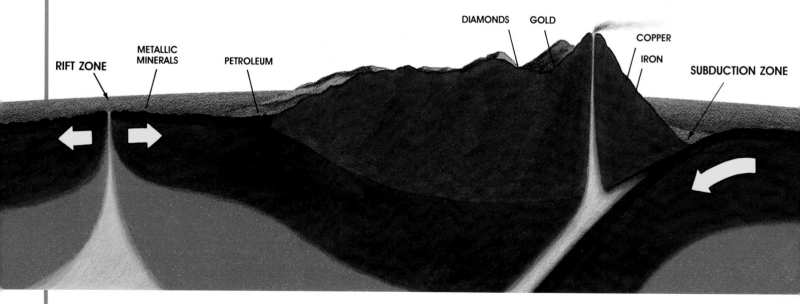

RIFT ZONE • METALLIC MINERALS • PETROLEUM • DIAMONDS • GOLD • COPPER • IRON • SUBDUCTION ZONE

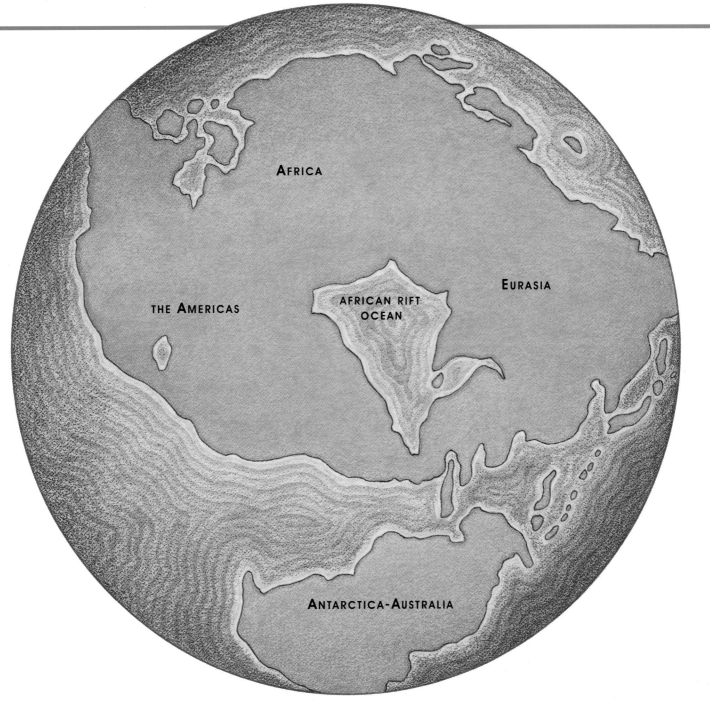

AFRICA

THE AMERICAS

AFRICAN RIFT OCEAN

EURASIA

ANTARCTICA-AUSTRALIA

In the future, the plates will continue to move. Rifts will widen, and plates may join together in a new Pangaea. The illustration shows how today's continents may look 250 – 350 million years from now.

No one knows for sure what the future will hold, but we do know that tectonic plates will continue to influence us, along with every other living creature. They will form new mineral riches, mountain ranges, deserts, and seas. They will reshape the surface of Earth as they create new lithosphere at spreading zones and swallow older lithosphere at subduction zones. And they will continue moving ceaselessly about the surface, jostling one another for space on our patchwork planet.

FOR FURTHER READING

- Ballard, Robert D. *Exploring Our Living Planet.* Washington, D.C.: National Geographic Society, 1983.

- Bonatti, Enrico. "The Rifting of Continents." *Scientific American* (March 1987): 97–103.

- ———. "The Earth's Mantle Below the Oceans." *Scientific American* (March 1994): 44–51.

- Brimhall, George. "The Genesis of Ores." *Scientific American* (May 1991): 84–91.

- Brush, Stephen G. "Inside the Earth." *Natural History* (February 1984): 26–34.

- Colbert, Edwin H. *Wandering Lands and Animals.* New York: Dover Publications, Inc., 1973.

- Frohlich, Cliff. "Deep Earthquakes." *Scientific American* (January 1989): 48–55.

- Gore, Rick. "Our Restless Planet Earth." *National Geographic* (August 1985): 142–182.

- Hoffman, Kenneth A. "Ancient Magnetic Reversals: Clues to the Geodynamo." *Scientific American* (May 1988): 76–83.

- Johnston, Arch C. "A Major Earthquake Zone on the Mississippi." *Scientific American* (April 1982): 60–68.

- Johnston, Arch C., and Lisa R. Kanter. "Earthquakes in Stable Continental Crust." *Scientific American* (March 1990): 68–75.

- Kerr, Richard A. "Tracking the Wandering Poles of Ancient Earth." *Science* (April 10, 1987): 147–148.

- ———. "Taking the Pulse of the San Andreas Fault." *Science* (January 27, 1989): 478–479.

- McKenzie, D.P. "The Earth's Mantle." *Scientific American* (September 1983): 66–113.

- Meissner, Rolf. *The Continental Crust: A Geophysical Approach.* San Diego: Academic Press, Inc., 1986.

- Miller, Russell. *Continents in Collision.* Alexandria, Virginia: Time-Life Books, 1983.

- Monastersky, Richard. "Spinning the Supercontinent Cycle." *Science News* (June 3, 1989): 344–346.

- NASA Headquarters. *Global Geodynamics.* Washington, D.C.: U.S. Government Printing Office, 1982.

- Overbye, Dennis. "The Shape of Tomorrow." *Discover* (November 1982): 20–25.

- Rona, Peter A. "Metal Factories of the Deep Sea." *Natural History* (January 1988): 52–56.

- Stein, Ross, and Robert C. Bucknam. "Quake Replay in the Great Basin." *Natural History* (June 1986): 29–34.

- West, Susan. "Patchwork Earth." *Science 82* (June 1982): 46–52.

- York, Derek. "The Earliest History of the Earth." *Scientific American* (January 1993): 90–96.

- Yulsman, Tom. "Our Restless Earth." *Science Digest* (March 1986): 48–54.

INDEX

INDEX